AN ADDRESS

ON THE

PROPRIETY OF CONTINUING THE

State Geological Survey of California

DELIVERED BEFORE THE LEGISLATURE

AT SACRAMENTO, THURSDAY EVENING,

January 30th, 1868:

BY J. D. WHITNEY,

STATE GEOLOGIST.

SAN FRANCISCO:
TOWNE AND BACON.
1868.

AN ADDRESS

ON THE

PROPRIETY OF CONTINUING THE

State Geological Survey of California

DELIVERED BEFORE THE LEGISLATURE

AT SACRAMENTO, THURSDAY EVENING,

January 30th, 1868:

BY J. D. WHITNEY,

STATE GEOLOGIST.

SAN FRANCISCO:
TOWNE AND BACON.
1868.

ADDRESS.

Gentlemen of the Senate and Assembly:

For the fourth time I am summoned to appear before the honorable Legislature of the State of California to give an account of my stewardship; and, embarrassed as I am between the desire of saying something which shall both interest and amuse, and that of crowding as much of instructive matter as possible into the allotted hour, I feel the necessity of asking your attention to what I have to say as a matter of business, even if I should not succeed in investing a somewhat dry subject with the graces of elocutionary display. There are 125,600 good reasons (almighty good ones to those who worship the almighty dollar) why I should be heard, since those figures represent the amount which the State has already expended on the Geological Survey, and if the money has been misspent, it is your duty to see that no more goes the same way. In inviting the State Geologist, therefore, to speak for himself, you have taken the shortest and most direct method of getting at the exact truth in this matter; and that the truth should be got at, before decisive action is had, seems to me no more than is reasonable to ask. I believe that, without exception, whenever there has been any opposition to our work, it has come from those who have taken no pains to inform themselves as to its real nature, from those who have never set foot in our office to examine what was going on there, and who have thus been actuated by blind prejudice rather than by any real desire to economize; for, not to put too fine a point upon it, the whole question of the continuance of the survey is simply a question of economy; it is just this — does it pay, or does it not pay? There is, I imagine, hardly a man in the State who would venture to oppose the continuance of the survey on any other ground than that of economy.

Not a whisper of opposition has ever been raised against the work, which has quietly and effectively been going on for the last twenty years on our coast, at an annual cost ten times greater than that of the Geological Survey, but at the expense of the United States, and the object of which is to give to this State, and all who need it, a map of the line along which old Ocean thunders against our rocky shores. No; the money does not come directly out of our pockets, let the great work go on; we see its value and approve its progress. But the Geological Survey—that is quite another thing: getting information *gratis*, and paying for the same, are two essentially different institutions.

A bill has been, or is to be, introduced, I am told, abolishing the State Geologist and consigning him to the tomb of the Capulets. This seems to me decidedly a work of supererogation, since that unfortunate officer is, to speak metaphorically, already in the last gasp of dissolution. If the Act is to be of any value it must be passed in a hurry, or you will be hanging a man who has just died of starvation. The train can be stopped just as well, and with less damage to the property of the State which is on board, by putting no more wood under the boiler and shutting off steam gradually, as by putting a rail on the track and thus throwing the engine off and smashing things generally.

At the time of commencing the survey, I had the honor of delivering my inaugural address before the Legislature, and in this I gave, in a highly condensed form, a history of the development of the mineral resources of the United States with special reference to what had been accomplished in this behalf by the different State Geological Surveys. Having glanced at the condition of the mining interests of this country, as compared with those of other parts of the world, I stated what the Act by which this survey was authorized called for, and laid out the work on which we were about to enter as well as could be done by one bringing with him a large amount of experience gathered in other regions, although but little acquainted, from personal observation, with the new field on which he was about to enter.

The next year, having the honor of addressing the Legislature without being especially called on by that body to select any particular subject, I endeavored to give some idea of the nature of geological inquiries, in their broadest and most generally attractive direction, and to awaken an interest in our work by setting forth some of the most interesting results at which geologists have arrived during the few years just past, and in various parts of the world, making my lécture a sort of spelling book and dictionary of our future reports; but not letting go my hold on my audience until, as in duty bound, I had said a few words in regard to the advantages to be derived from a thorough prosecution of the work on which we had then fully entered.

Once more the State Geologist appeared in this place and addressed the Legislature, by request, on the question of the establishment of a State University in California, giving some of the results acquired by many years' experience as a pupil, resident graduate or professor in several institutions of learning of the highest rank, on both sides of the Atlantic, including Yale and Harvard Colleges, the School of Mines in Paris and the Universities of Giessen and Berlin.

On both these last occasions I endeavored to set forth in as clear a light as possible the relations of the survey to the cause of higher education in this State and on the Pacific Coast in general—a subject which is by no means exhausted, and on which I will ask permission, before closing, to add a few more last words, in view of the fact that the question of a State University is again before you for discussion, and in the hope that my ideas may be found not unworthy of being heeded.

On this solemn occasion, when the fate of the survey seems about to be decided, and I, perhaps, may be preaching its funeral sermon, if I were to search for a text on which to base my discourse, I do not know that I could find one better than that containing a malediction on the man who looks back after putting his hand to the plow; for this text exactly expresses the sentiment which has actuated me in pushing this great work along against every obstacle — want of sympathy among the people, want of sufficient appropriations from the Legislature, want of sufficient knowledge on the part of many to understand the real extent and probable value of our results, but no want of misapprehension and misstatement of our motives and actions, or of abuse in the newspapers for not doing what we have done, and for doing that which we have not done. A constitutional antipathy on the part of its chief to looking back after having laid his hand on the plow is the one effective reason why this survey has not long since been wound up and its fossil remains left in the pigeon-hole, over which, in big black letters, the ominous word "Fizzle" stands, as representing a great undertaking abandoned for want of pluck and energy to see it through.

It has been repeatedly thrown in our teeth that we were trying to accomplish too much; that our plans were too extensive, and that we were aiming at something it was beyond our power or that of the State to accomplish. Let us see: the Legislature ordered "An accurate and complete geological survey of this State, with proper maps and diagrams thereof, together with a full and scientific description of its rocks, fossils, soils and minerals, and of its botanical and zoölogical productions." I quote the exact language of the Act. Suppose, now, it had read thus: "The State Geologist shall make a hasty and inaccurate survey of the State, and furnish unreliable and worthless maps of the same, together with a popular and amusing account of his travels, and private reports to mining speculators, on the principle of the bigger the fee the more favorable the report." Does any one suppose that a scientific man, with a reputation and a conscience, could have been found to lend his name to such a ridiculous proposition? And the scientific man who should make himself responsible for the statement that anything but a hasty and inaccurate, and consequently worthless, survey could be made without much time, labor and money, would be either a knave or a fool, or both.

The State Geologist is not responsible for the plan of the survey; all he has sought to do was to carry out the arduous task set before him by the Legislature to the best of his ability, giving the best years of his life and his undivided energies to the work, with no other object in view than that of so executing it that it would be of permanent value to the State, and consequently a credit to himself.

What, then, is the object of the Geological Survey, and for what purpose has it been instituted? This question I will endeavor to answer, and I will then show, as far as possible, within the limitations imposed on me by time and place, how much progress has been made in it; and finally, will give some reasons why, as I think, the work should be continued to completion, and that on the scale and with the plan on which it was started, and on which it has thus far been carried on.

The object of the Geological Survey may be succinctly stated in these words: "It is to give to the world, and especially to our own citizens, an encyclopædic statement of the natural resources and capabilities of the State." Its scope may, perhaps, be better comprehended if we consider what a private individual would do if he were to come into possession, by legacy or otherwise, of a vast estate of unexplored and unsurveyed territory, in regard to the value of which there were no certain data in existence. If unable to examine his newly acquired property himself, he would hire others to explore, survey and map it, to investigate its capabilities for settlement and its resources for sustaining a population, so that he might be able to cut it up and bring it into the market, or otherwise to make it available. This is just exactly what we are doing for the State ; we map its surface, examine and describe its natural productions, whether animal, vegetable or mineral, and do, in a large way, as a unit, for the State and all the inhabitants in it, just what each individual would wish to have done, if he had intelligence enough to know what was for his own interest and the means to accomplish it, for his share of the great estate which belongs to the people and is to be made available for their benefit, individually and collectively.

The object of the Geological Survey being, as has been stated, to furnish an encyclopædia of the resources of the State, the mode in which this object was to be best accomplished, taking into view on the one hand the needs, and on the other the resources available, gradually shaped itself in the mind of the individual to whom the work was intrusted.

As thus developed, the survey is divided into three principal departments, each again with its subordinate divisions.

The three main divisions are : Geography, Geology, and Natural History. The first includes a topographical survey, the publication of maps, and also an account of the physical geography of the State. The second department includes general geology, economical geology, and palæontology. The third, botany and zoölogy. Thus, there are seven subdivisions of our work, each requiring one or more volumes for its complete illustration.

I will now proceed to state, as briefly as is consistent with intelligibility, what has been done under each of these heads, aiming to give some general idea of what we aspire to accomplish, and setting forth what remains to be done as the supplement of what has been done. At the foundation of our work lies the topographical survey, for without a map of the State we should be as much at a loss to describe its resources, as a painter would be in working without a canvas on which to embody his conceptions. It is not necessary, and would not be even before the most illiterate audience, to enlarge on the necessity and importance of geographical maps to every country ; as well might I undertake to demonstrate the desirability of putting up the frame of a house before putting on the clapboards and shingles. The exact stage of civilization of every country or State can at once be inferred from the character of its maps.

But a complete and accurate survey and map is necessarily a work of much time and labor. To give an idea of what has been expended in this way in other parts of the world, I will state a few facts gathered from the United States Coast Survey Report for 1858.

The total area of the United Kingdom of Great Britain and Ireland is about 120,000 square miles, or 40,000 less than that of California. The entire cost of the topographical surveys of that country had been, up to 1854, $12,000,000. And to complete the work, $20,000,000 more was required; 3,500 persons were engaged on the survey at one time. This is exclusive of what was being spent at the same time on the geological and hydrographical surveys.

The survey of France — an empire about one-fifth larger than our State—was commenced in 1818, on an estimate of thirty years for the time required to complete it and an expense of $20,000,000; 2,500 men, besides laborers, have been employed in the work.

Massachusetts, with an area only one-twentieth that of California, spent more on her geographical map than our whole survey has cost, and that more than thirty years ago.

It may be said, however, what is the use of parading these figures, which seem to demonstrate that the possession of a good map is something that we cannot aspire to? To this, I reply, in the first place, that we are doing this work on a very limited scale, as regards expenditure, and that we shall accomplish a great deal with a comparatively small amount of money; next, that if it is not done by the State it will be attempted to be done by individuals, and the result will be that no good map will ever be obtained; while, in reality, a much greater expenditure of money will be made. Private parties will be continually getting out new maps, each one of which will be a little more of an approximation to the truth, but at the best extremely defective; while the public will be continually buying these maps as they appear, rejecting the old ones already on hand, in the vain hope that each new one will be sufficient for their needs—a system which will be equivalent to laying the people of the State under a perpetual tax for the benefit of a few map makers, who work without ever having any prospect of attaining satisfactory results. Those who have not examined into the matter have little idea how large sums are spent in this way in the State. I consider myself safe in stating, that an amount greater than the entire cost of the survey has been paid out, during the time our work has been going on, for the imperfect maps which have been issued. And this state of things will go on indefinitely, unless put a stop to by the State by continuing our work to completion; because it is for the interest of the parties engaged in this business to have an excuse for issuing new maps as often as possible, just as milliners and dealers in dry goods arrange their business so that the fashions may change every three months at least, ostensibly for the benefit of the ladies, but really for their own, and to the great detriment of the unfortunate husbands.

It may be said, also, that the United States surveys will give us, eventually, a correct map of the State, and that it is therefore unnecessary for us to do it: this is not the case, for a most careful examination of the United States work shows clearly that it can never, in a mountainous country like California, be coaxed into anything like a permanently valuable map. The town and section lines are run in the valleys, it is true, and were this State a vast plain, these lines would give us a general idea of the country, as they have in the great Mississippi Valley; but, in a region like our State, of which less than one-fifth of the surface is plain or valley, they are of no account at all, especially as topography is no part of the idea of the Government in having the lines run, while the work itself (most of it, at least) is so carelessly and even fraudulently done that it is impossible to make it fit together. In making accurate surveys of regions where the town and section lines have been run by the Government, we have found sometimes that a line supposed to be a mile in length, and measured as such in the United States linear survey, was in reality a mile and three-quarters long, so that the net-work of Government lines, when laid down on the paper as they actually are, and not as they profess to be, look somewhat as a gridiron struck by lightning might be supposed to. This need not always be the fault of the surveyor, as the system itself is one that is not in the slightest degree applicable to the survey of mountainous countries. In the southern part of this State millions of dollars have been paid for surveys which were in reality never executed, as we find, when we go over the ground, that there is not the least resemblance between the topography as laid down on the official maps and that which our work shows it to be.

Our plan of operations and publication has been carefully adjusted to meet the wants and the means of the State. We propose to publish maps on different scales, all accurate as far as they go, but, of course, with a varying amount of detail, to suit the condition of different sections, basing the amount of detail on the density of the population of the section mapped. For the whole State we take a scale of ten or twelve miles to the inch, which will give us a map about five feet square, and as large a one as can conveniently be used for a wall map for schools and for the people at large. For the central portion of the State we take a scale of six miles to the inch, which gives us four times the area of the other. This central map embraces only one-third the area of the State, but it includes over ninety per cent. of the population. This map is well under way, the field-work being about four-fifths and the drawing one-half done. It can be completed entirely in the next two years, with a reasonable appropriation, and when done will be the largest inland piece of map-work yet undertaken in the country, as it will give the details of the topography of 80,000 square miles of territory—an area nearly twice that of Ohio. The same scale is adopted for the Coast Ranges south of Monterey as far as Los Angeles, and this map is about two-thirds completed.

For the most thickly settled parts of the State we have adopted a much larger scale, of two miles to an inch namely, giving an area of nine times that of the last mentioned map for the same territory. Of the work done on this scale you have before you a sample, which will render it unnecessary for me to go into any details in regard to it, and which will enable every man to judge for himself of the value of the survey maps. The one in question is the Map of the Vicinity of the Bay of San Francisco, of which a large supply is now on the way from New York, in different styles of mounting.

Of the belt of mining counties along the Sierra Nevada, three maps on this scale are in preparation: one, that of Plumas and Sierra, is done as to the field-work, and the drawing of it will be completed during the winter, so that it can be engraved next summer if our work goes on. The central counties, from Nevada to Calaveras, are also well under way, and the southern begun; the rate of progress will depend, of course, on the amount of funds provided by this Legislature for the continuance of our work.

According to my calculations the whole of the map-work can be completed in four years, if pushed with vigor, and I consider that, taking all things into consideration, it may be considered now as nearly half done. The question, therefore, before you is, not whether a topographical survey of the State shall be made; but whether, one having been commenced on the authority of one and continued on that of four successive Legislatures until nearly half done, it shall be abandoned just as its results are beginning to be laid before the people.

It is intended that all the maps published by the Survey shall be sold singly, mounted according to the fancy of the purchaser, or in plain sheets, and that they shall all, at the close of the work, be collected and bound into a volume forming one of the series of our Report. I do not hesitate to say that they will form a series of which the State may be proud, and which will be considered by persons acquainted with such matters as fully repaying the entire cost of the survey. With the aid of our maps, each county can, by the help of the County Surveyor, and at a comparatively trifling expense, have a special county map of its own, on which such items may be inserted as are peculiarly desirable for county purposes, and which can be taken from the official records with the sanction of the Supervisors.

The geographical discoveries of the Survey in this State have been of great interest, having brought to light much that was new and curious in regard to the peaks, passes, mountains, and valleys of the Coast Ranges and the Sierra. We have opened a new region to the traveller and the tourist, as large as Switzerland, of which the mountain peaks surpass those of the Alps in elevation, and which in grandeur of scenery is without a parallel on the continent. If this region had ever been explored or visited by any one before us, no record exists of such exploration, nor had ever one word been written or published in regard to it, until the Geological Survey made its existence known. And let not the importance of such discoveries be under-estimated. Few persons who have not turned their thoughts in that direction, with some knowledge of what is going on in other parts of the world, have any proper idea of the real value to the State, in a pecuniary point of view, of its natural scenery. Superficial observers may not recognize the fact that the picturesque is an element in the resources of the State as much, if not in as great a degree, as its agricultural and mineral capacity. The time will come when the money brought into this State by pleasure travellers will be, if not as important an element in our prosperity as it is in Switzerland, at least no mean addition to our resources. By opening up our grand scenery, describing and mapping our most picturesque regions, and spreading a knowledge of them through the world, the survey has done the State a great pecuniary service, which will be recognized in the future if it is not now.

The results obtained in the department of Physical Geography, such as the elevations of towns, mining camps, valleys, mountains and passes, the distribution and character of animal life, forest and plant vegetation, climatological data, circumstances bearing on agricultural capacity, and many other points of this kind, are all of great value, not only scientifically, but practically. The records preserved in the office of the survey are constantly being consulted by those who are seeking information in regard to all kinds of public improvements on this coast; and if one-half those who have thus been benefited by our work could appear before you and give their testimony, I have little doubt but that it would furnish an overwhelming mass of evidence in our favor.

In the geological department of the survey, with its subdivisions of general geology, economical geology, and palæontology, good progress has been made, and the way prepared to make a much more rapid advancement in the future, if we have the necessary pecuniary assistance. It seems almost an absurdity, at this late day, to be arguing in favor of a geological examination of a great and little known mining country like that of California. The fact that such surveys have been made in all civilized foreign countries, and in nearly every one of the United States, in many of which this kind of investigations had not one-tenth part of the interest which they have here, is sufficient *prima facie* evidence in favor of their value. The fact, also, that five successive Legislatures have, after due investigation, given their verdict on the importance of this survey, seems to me to offer a very fair guarantee that it was undertaken with sufficient reason, and the only question which ought legitimately to come up, at the present time, would be: Is the work being carried on in a satisfactory manner?

As, however, each Legislature has the power of upsetting that which has been done or begun by its predecessors; and as, therefore, the original question of the propriety of such a work as ours must be started anew and discussed at each session, regardless of all previous indorsements, public or private, I will take the liberty of stating once more what the special aim of the geological part of the survey is, before undertaking to give an account of the progress made in that department of our work.

The strictly geological portions of the survey may be divided into two sections. The first includes the general geology and palæontology; the second, the economical, or applied, geology. Under the first division we include all that relates to the general geological structure of the State, while the second embraces the practical application of the information thus obtained to the wants and uses of the people in the arts, manufactures and commerce. Under the head of general geology, we have to investigate the nature of the different rock formations which are spread out in the valleys or piled up in the immense mountain masses which traverse the State. We endeavor to ascertain of what materials they are composed; how originally formed or deposited; what changes they have undergone since their deposition, and by what agencies these changes have been brought about. We search for and describe the fossil remains which the stratified rocks contain, and thus are enabled to compare them with the formations of other countries and to fix their relation, geological age and position. We then trace over the surface of the State and lay down upon the map the range and extent or the geological distribution of the different systems and groups of rocks, and exhibit their stratigraphical relations, or position with regard to each other, by means of sections, showing the configuration of the surface and the character of the rocks beneath it, along certain lines measured and examined for the purpose. By these preliminary operations we are prepared with the necessary basis on which to proceed with the next great division of our work, namely, the economical geology. Without this scientific part of the survey, the practical would have no permanent value, for it would be nothing but an incoherent mass of material such as our newspapers are filled with, not put into form or reduced to system, so as to be generally applicable and easily comprehended. After and while tracing out the various geological formations and getting their sequence thoroughly established, we endeavor to discover and classify the metallic and mineral treasures which they contain, to ascertain their position and mode of occurrence, or, in other words, to gather all the facts necessary to enable us to determine their present and prospective value, and to show how and under what conditions they may be best made available for the industrial purposes of life. In doing this we furnish a basis for detailed explorations for further deposits of metallic and mineral treasures, by limiting the field for research for numerous prospectors always engaged in the search for useful ores, so that every man will be working where his labor will tell, and not throwing it away in undertakings which a comprehensive view of the mode of occurrence and geological position of our economically valuable materials will show to be a mere waste of money, time and energy. I do not hesitate to say that millions have been wasted in this State, for want of just that information which we shall be prepared to supply, and which, indeed, we have already supplied to a considerable extent to those asking for it. Every year sees an addition to the number of persons availing themselves of information, which we are always ready to give, on matters connected with the geological mode of occurrence and the probable value of deposits of all kinds of metal and mineral; and I know that, in some cases, our advice has been followed with manifest advantage, and that, as time goes on and demonstrates the reliability of our work, and the substantiality of the basis on which our opinions are founded, the survey will be more and more efficient as a break on the wheel of reckless mining expenditures. If this survey could have been begun at an early period in the development of the State, and have firmly established itself in the confidence of the people at the time of the last great mining excitement on this coast, of what incalculable value might it not have been! I am aware that there are persons so little acquainted with the principles of political economy and the laws which govern the progress of nations, as to think that money expended in the State is a benefit to it, whether any results of permanent value be attained or not by such expenditure. Benjamin Franklin had a clearer idea of the truth when he put into the mouth of Poor Richard his famous saying: "A penny saved is a penny earned." It is only in communities where the pence are saved that the great results of a permanent and high civilization are obtained. To insure permanent working and economical development of what is discovered, by giving every one the means of knowing beforehand how his discoveries may be turned to the best account, how he can best open his mine, how treat his ores, what form to give his products and where and in what quantity they can be disposed of—these are some of the more important points to be gained by the development of that department of our work which is included under the designation of Economical Geology. The services which we shall be able to render, in this line, will become every day more important, as our basis of experience is enlarged and as it becomes more clearly understood that our opinions are disinterested, and that we have no other objects in view than the welfare of the State and the development of its mineral resources. In our volumes devoted to Economical Geology we shall throw all possible light on these subjects, and it will not be our fault if the man about to embark in any undertaking connected with ores or mineral substances shall not find in our book something which shall materially aid him in his undertaking, or at least prevent a foolish waste of money on the impracticable. It is, in every respect, desirable also that the resources of the State should be made known to the world, under official guarantee of correctness, so that not only our own capitalists, but those of other countries, should have opened to them a field for investment, in regard to which they may understand that they have some substantial basis of facts and reliable data for generalization, and not feel that they are entering blindfolded into a hap-hazard game of speculation, as is too often the case when putting their money into mines in regions known only from the descriptions of those personally and pecuniarily interested.

In the geological department proper of the survey, one volume has already been published, which, under the title of "A Report of Progress and Synopsis of the Field-work from 1861 to 1864, inclusive," gives the results of a general reconnaissance of the State, both geological and geographical. In this volume the main features of our physical geography and geology will be found delineated, exclusively from the results of our own observations, and with them is incorporated a considerable amount of general information with regard to our mineral resources, incidentally brought in, as also notices of our natural scenery, botany, distribution of forests, etc., all of which subjects will eventually be more elaborately treated in special volumes; so that this might be considered rather in the light of a temporary report than as a part of a final series. To close the general geology another volume will be required, and is in process of preparation. This will give a systematic and thorough review of the geological structure of the State, and will be fully illustrated by geological maps and sections, which were necessarily wanting in the first volume, and which will be a text-book for the student in this department, and a reliable guide to those who, from whatever motive, shall desire to make themselves acquainted with the course of events which has given its present configuration and internal structure to that part of the continent which we inhabit.

In Palæontology, we have published one volume, and another is well on the way, a considerable part of the plates being already engraved and the text stereotyped. A third volume will be necessary to enable us to describe the remarkable animals which lived on this coast just before the present epoch, including the elephant, mastodon, camel, tapir, horse (of several species now extinct), buffalo, rhinoceros, hippopotamus, and others of remarkable and little-known genera; also, the forest vegetation of that epoch, of which the remains now lie buried in our deep gravel deposits, and which differed entirely from that which we now see occupying the flanks of the Sierra; also, the microscopic organisms of which a large portion of our rocks are almost entirely made up, and which can be shown to have a very important bearing in an economical point of view, as constituting the origin and source of the bituminous matter, asphaltum and petroleum, so widely distributed over the State.

In the department of Economical Geology, less progress has been made than would be desirable, partly on account of the necessity of its being preceded by the other departments, in order that it may have a safe basis on which to stand ; but more because the appropriations have been insufficient. It stands to reason that the necessary assistance to thoroughly work up this department cannot be had without paying for it. Men qualified to do this work can obtain large salaries as heads of mines and mills ; and, if asked to give their services to the State for half or quarter of what they can obtain elsewhere, are very apt to see it in another light than that of exclusive devotion to science regardless of personal considerations. And if I should take young men and educate them until they became competent for the work, I very much fear that the result would recall to mind the modern reading of an old text: "Train up a child and away he goes."

The first volume in this department is, however, well under way and can soon be put to press ; it will contain the non-metalliferous minerals occurring in the State, that is to say, all materials used as fuel, for illuminating, or for building purposes, including coal, asphaltum, petroleum, lime, cement, plaster, marble, granite, and the whole long list of substances of mineral origin used in the arts, and not in the metallic form. A most careful examination has been made of all localities where bituminous materials of any kind occur in considerable quantity ; specimens have been collected and subjected to chemical analysis ; new processes have been contrived for making them available, and the results, when fully reported, cannot fail to interest all who are turning their attention in the direction of available fuel and illuminating materials. The coal interests of this coast are also of great importance ; they require and have received a large share of attention.

We come next to the Natural History department of the survey. This is divided into botany and zoölogy, as before stated, having for its object a complete description of all the forms of animal and vegetable life occurring naturally within our borders.

The department of Botany was under the direction of Professor Brewer, now of Yale College, while he was in this State, and is now, so far as the working up of the flowering plants and publication of the results obtained is concerned. Ever since the commencement of the Survey, a vigorous collection of materials has been going on, first under the direction of Professor Brewer, and now of Mr. Bolander. This collecting, carried on in all quarters of the State, has furnished a vast mass of materials, including a great number of entirely new genera and species, and these have been distributed to the most eminent botanists in the country to be worked up—an operation which has been going on steadily for the last three or four years—the work being so far advanced towards completion that it is thought that the volume of flowering plants may be put to press towards the close of the present year ; but that, at all events, it can be finished and printed in time for delivery to the next Legislature.

In Zoölogy there are four volumes under way, and in different stages of preparation, the text of all being well advanced, and nothing required to enable them to be put to press excepting the completion of the illustrations, the drawing and engraving of which has been going on for more than three years.

Having now given a rapid sketch of the general progress of our work, and shown something of the results accomplished in each department, and of what more is proposed to be done, provided we are furnished with the means, I will proceed to discuss a few points connected with the existence and completion of the Survey, a little more in detail than I have been able to do in the preceding systematic review.

Assuming that the plan of the Survey is a good one, and judiciously contrived, the question arises—how is it being carried out? Forming a plan is one thing, and executing it another. To this question I can only reply that our woık, as far as accomplished, has been submitted to those who would unanimously, among scientific men, be regarded as best qualified to judge of such matters, and has met with their warm and decided approval. I have assumed that if the survey was done in such a manner as to win the applause of the highest authority in science, in this country, that I might consider it as being well done. And when I say highest authorities, I mean such men as Professor Bache, the late eminent Superintendent of the Coast Survey, Dana of Yale College, Agassiz and Gray, of Harvard, Henry and Baird of the Smithsonian, Lea and Leidy of Philadelphia, Guyot of Princeton, etc. From all of these, letters are on file at our office, expressing sentiments of the warmest interest and the most entire satisfaction in regard to our work, and which are at the service of any member of the Legislature to read and examine. I will take the liberty of reading part of one myself. It is from one of the heads of the Smithsonian Institution, and addressed to the State Geologist, under date of October 18, 1865:

"Volume I of the Geological Report of California is a work of which the State may well be proud, as, while of almost unrivaled typographical execution, its contents are of the first order of scientific merit. It needs but the full completion of your plans in regard to the entire series to give to the Pacific slope of the United States an encyclopædia of information respecting its natural and physical history far more perfect and complete than is possessed by any other State in the Union, New York even not excepted. You may safely assure the Governor and Legislature of California—if such indorsement is necessary—that there are no dissentient views among the men of science here as regards their interest in the Survey in its various branches, and their satisfaction with the character of its plans and execution, as far as it has gone."

To this it might, however, be answered by the opponents of the Survey that a jury of scientific men is not the proper kind of a one to sit in judgment on this work. If that be the case, then I am not the proper kind of a person to carry on the Survey; and to adopt such a principle, or suggest any other tribunal than a scientific one, would be at once to destroy all that gives character and respectability to our work. We assume that those men who have devoted their whole lives to investigations of this kind, and attained the highest positions and universal recognition as representative men in science, are best qualified to judge in regard to the value of work done in their respective departments, and that if the authority of their opinions is not appreciated by the people at large, it is because the people have not arrived at a sufficiently high stage of educational development to understand what is for their own interests. But I do not mean to be understood as saying that non-professional men have not given us their hearty support in many cases, and that the Survey is only appreciated by the few. On the contrary, we have received the most satisfactory assurances of sympathy and regard for our work, and of its practical value, from many who would not claim to be considered as other than practical men themselves. The only difficulty, as before hinted, has been to induce the opponents of the Survey to give our work a candid examination, or any examination at all. They have conceived a blind prejudice, based on some little matters which have no relation to the real merits or plan of the survey, and have acted accordingly, entirely regardless of the fact that, if successful in their opposition, they would be incurring an amount of odium which every year's advance of the State toward a higher plane of civilization would not fail to increase.

The question of the establishment of a State University is again before the Legislature of California, and this time in a more tangible form than ever before. Indeed it is understood that a site has been selected, and there seems to be a general calling from all quarters for some positive action which shall set the wheels in motion. In view of these facts, I deem it more than ever justifiable in me to call attention to the fact that the Geological Survey is a necessary preliminary to the establishment of a University which can claim to be anything more than a name.

It is not from the standpoint of a Professor, or from any supposed right to be heard, based on an intimate acquaintance with the organization and management of several of our higher institutions of learning, including both those connected with the various State Governments, and those independent of them—those counting the years of their existence by hundreds, and the amount of their endowment by millions, and those whose career has but just begun, and who are proportionately short of funds—it is not, I say, on any such grounds as these that I approach the subject; but simply as one called before you to defend the Geological Survey, and who desires, as one of the important points in this defense, to urge upon you the educational relations of our proposed work, not only as connected with the proposed University, but with all our schools and institutions of learning. For I take it that there is no institution of so low a grade that the leading facts of the geography of the State should not be taught in it, and that we should not have to rise very high in order to come to those in which instruction should be given in the elements, at least, of natural history. But it is to the proposed University that I especially refer, as our work is more intimately connected with that than with institutions of a lower grade.

In a University established under the conditions which surround us on the Pacific coast, it is not difficult to see that the practical will have very much the upper hand over the theoretical and abstruse; that modern languages will outweigh the ancient, and that the natural and physical sciences will be more cultivated than psychology and metaphysics. There will be little call for Latin and less for Greek; but nature will be interrogated, and everything that aids in familiarizing the student with her teachings will be in demand. The scientific branches in which instruction will be most craved by the student, will certainly be physical geography, geology, mineralogy, botany and zoölogy; these at least will be departments which as much if not more than any other it will be necessary to have filled, if the proposed institution is to have any rank at all, or to be in any respect up to the standard of other colleges, universities or schools of science. Make the institution as practical as you please, lower its grade to the last conceivable degree, still the great fact cannot be got over that something has to be taught there; that there must be some course of study, and that whether you simplify or complicate the programme, the already mentioned branches will be the last to be omitted from it.

Now I state what I know to be a fact, when I assert that no one of the branches in question can be taught in any other than the most superficial way until the results of our Survey are in the hands of the teachers. This statement I will illustrate by reference to our botanical work, as this may probably be better understood by the people generally than any of the other departments mentioned. It will probably be admitted by all that it would be absolutely necessary for the teacher of botany, in order to impart to the student anything more than the merest rudiments of the science, to know the names and position in the system of the plants which grow in his vicinity, and which he would collect and use for illustration of his teachings, and which his pupils would bring to him for that purpose. This knowledge, however, is an impossibility at present; there is no botanist in the State who can name the plants he collects, nor is there any one person elsewhere who can do it for him.

The facts are simply these: During the last seventy years, more than one hundred and twenty professional botanists and collectors have visited parts of the regions west of the Rocky Mountains, and more than seventy of them have traveled in California. Their collections have gone into the various herbaria of this country and of Europe, and the printed data relating to them are scattered through an immense number of volumes; so that if any one were to begin and make a complete collection of them, it would require years of labor and thousands of dollars expenditure, and, even with the most strenuous exertions, it would be impossible to make the list complete. But if this could be done, and this library could be transferred to the Pacific coast and placed in their possession, our botanists would still be unable to give correct names to their specimens. And for these reasons: not unfrequently several collectors have obtained the same species in different localities and seasons and in varying forms; specimens thus collected have been referred to different botanists for description, the amount of material being often meager and insufficient, and the results have been published at places widely remote from each other. Thus, what was in reality one and the same species, has often several names attached to it; but to discover this fact and clear up all the difficulties, so that all the synonyms should be properly arranged under the real name, or the one having priority, according to the universally recognized rules of scientific nomenclature, requires not only the extensive collections of the Survey, made under the most favorable circumstances, in all kinds of localities, and under all conditions of growth, but also an actual inspection and diligent study of the original specimens collected by all botanists prior to our work. These are, luckily, mostly accessible in the grand herbarium at Cambridge, which Professor Gray, the leading botanist of the country, has for the past forty years been gathering together. Without the aid of this gentleman it would be impossible for any one to get through with the immense undertaking of bringing order into the chaos of California botany. And not only has he kindly lent us the use of his herbarium and library, and given his personal attention to the description of the great number of new plants collected by our parties, but the same may be said of every other eminent botanist in this country: Engelmann, Torrey, Eaton, Tuckerman, Lesquereux and Thurber, as well as several of the most distinguished authorities in Europe, have lent us a hand in unraveling the twisted skein, and one gentleman is about to leave his home in a western city and visit Europe for the purpose of comparing specimens of the California pines and oaks with the authentically named ones existing in the herbaria in England and on the Continent. I should add, lest anybody's sensibilities be alarmed by the excessive expenditure, that he receives no salary, and travels at his own expense.

We shall thus have a work, in the botanical department, in which each plant, in every important group of families, will be authentically named by the highest authority in that section of the science—a book which every student can use with perfect confidence in its reliability, and which will be the indispensable guide of every teacher of the science. And we could not have had it in any other way than this: it required a combined effort of all the botanists in the country, sanctioned by the State, to do this work; and with all the facilities the survey affords, this task is an arduous one.

The same thing may be said (*mutatis mutandis*) with regard to the other scientific branches mentioned. I will not tire you by going through them all, but will merely add that great pains have been taken with the illustrations of the natural history volumes to make them such as will, while possessing a high scientific value, be most useful to students. And at the same time economy has been studied, so that I am fully justified in assuring you that, while our series will be more complete than those of any other State, they will also have cost much less.

In fact, this idea has been constantly in my mind, while engaged in the Survey, that our work was the necessary preliminary of the University and the cultivation of science on this coast in general, and I have endeavored so to shape our plans that when our task has been completed the way shall be smoothed for others to carry on that which we have begun; for so inexhaustible on this is nature in her ways and works, that we cannot look forward to any time when the student can fold his hands for want of something to do, in the way of enlarging the boundaries of either natural or physical science.

I need not enlarge on the importance of our full and authentically named collections, in the various departments, to the State University. They will form the foundation of a museum invaluable for the purposes of instruction, and such a one as could not have been brought together without a thorough and systematic plan of operations. The collections are amply sufficient to justify their being divided, and I trust that when they pass out of our hands they may be made as available as possible, both for scientific and popular instruction. This can be best accomplished by giving one series to the University, and the other to the Academy of Sciences, in San Francisco, in which latter place a museum will be accessible to a much larger number of persons than it would be anywhere else, and where our materials would be added to a large and rapidly increasing collection, in charge of the only scientific association on the coast, and which, in time and with the growth of a liberal and scientific spirit in this State, will have the means to display them in a suitable manner and to preserve them from destruction.

Finally, I conceive that the Survey ought to be carried on because it has been begun—not only because it has been indorsed by several successive Legislatures, and by eminent scientific authority at home and abroad, but that the State may not be making an exhibition of herself, in the eyes of the world, as a specimen of fickleness and unreliability. Having published far and wide that we were going to have a thorough geological survey, and having been liberally patted on the back for our energy and far-sightedness, I can conceive of nothing more humiliating than backing down, without reason, when the work is already more than half completed, and the most practically valuable portion of the results is just beginning to see the light.

I am aware that there are some who would pull the survey up by the roots in order that the University may be planted in the same hole, with the idea that there would be economy in that operation, or else with some hidden notion that does not appear on the surface. There might have been some reason in this plan of making the Survey by the University, or a University out of the Survey, for the two things amount to about the same, had it been put in operation at the commencement of our work. Indeed, something like this was suggested by the State Geologist, three years ago, in an official report. It is now too late: as organized, the Survey requires the entire thoughts and time of every man connected with it, and there has never been a new College or University established in the country where the Professors had not their hands full in attending to their legitimate official duties, and few is the number of them who are fitted by education or practice to engage in a work such as we are carrying on, without special preliminary training. To stop the Survey in order to encourage the University would be like pulling the foundation of a building to pieces in order to get material for the walls and roof.

By what I have said here to-night, I am aware that I have laid myself open to the charge of blowing my own trumpet; but in what I have said I am very sure that I have not exceeded the truth, and that I have not given an opinion not backed by many years of experience. I cannot refrain from adding, however, that I have carried on this Survey in spite of many obstacles and great temptations to engage in other less laborious and responsible, but more lucrative, work. Enough has been accomplished to show to the world what it would be were my plans to be carried out, and thus to take from my shoulders the responsibility of the failure if the Legislature chooses to bring the work to an end. If the Survey can be continued on the same basis on which it has thus far been prosecuted, free from all political contaminations, and with the same ideal of thoroughness, and with a sufficient liberal appropriation to insure a rapid carrying on of the work, I shall rejoice to go on with it and complete it, as I fully believe that it will reflect the highest credit on the State, and all officially connected with it, as well as the Legislatures by which it has been upheld. Let the work stop, without any fault or *laches* of mine, and I shall feel that it is not of me that it can be said that, "having put his hand to the plow, he looked back."

www.ingramcontent.com/pod-product-compliance
Lightning Source LLC
LaVergne TN
LVHW011140110826
845150LV00008B/2435
* 9 7 8 1 4 1 8 1 9 2 7 1 6 *